Elements

ALUMINIUM

Al

Atlantic Europe Publishing

How to use this book

This book has been carefully developed to help you understand the chemistry of the elements. In it you will find a systematic and comprehensive coverage of the basic qualities of each element. Each two-page entry contains information at various levels of technical content and language, along with definitions of useful technical terms, as shown in the thumbnail diagram to the right. There is a comprehensive glossary of technical terms at the back of the book, along with an extensive index, key facts, an explanation of the Periodic Table, and a description of how to interpret chemical equations.

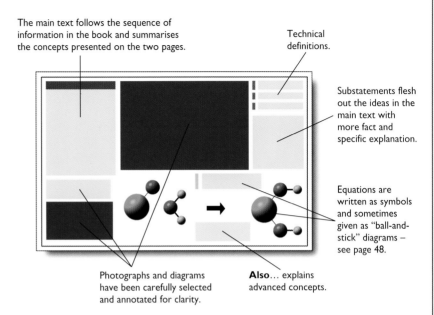

The main text follows the sequence of information in the book and summarises the concepts presented on the two pages.

Technical definitions.

Substatements flesh out the ideas in the main text with more fact and specific explanation.

Equations are written as symbols and sometimes given as "ball-and-stick" diagrams – see page 48.

Photographs and diagrams have been carefully selected and annotated for clarity.

Also… explains advanced concepts.

- -

An Atlantic Europe Publishing Book

Author
Brian Knapp, BSc, PhD
Project consultant
Keith B. Walshaw, MA, BSc, DPhil
(Head of Chemistry, Leighton Park School)
Industrial consultant
Jack Brettle, BSc, PhD (Chief Research Scientist, Pilkington plc)
Art Director
Duncan McCrae, BSc
Editor
Elizabeth Walker, BA
Special photography
Ian Gledhill
Illustrations
David Woodroffe
Electronic page make-up
Julie James Graphic Design
Designed and produced by
EARTHSCAPE EDITIONS
Print consultants
Landmark Production Consultants Ltd
Reproduced by
Leo Reprographics
Printed and bound by
Paramount Printing Company Ltd

First published in 1996 by
Atlantic Europe Publishing Company Limited, Greys Court Farm,
Greys Court, Henley-on-Thames, Oxon, RG9 4PG, UK.

Suggested cataloguing location
Knapp, Brian
    Aluminium
    ISBN 1 869860 44 6
    – *Elements* series
540

Acknowledgements
The publishers would like to thank the following for their kind help and advice: *Alcan International, British Alcan Aluminium plc, Kjc Operating Company, Dr Angus W. R. McCrae, Rolls-Royce plc, Frank Sperling* and *Pippa Trounce.*

Picture credits
All photographs are from the **Earthscape Editions** photolibrary except the following:
(c=centre t=top b=bottom l=left r=right)
courtesy of **British Alcan Aluminium plc** BACK COVER, 4/5c, 10b, 10/11t, 12t, 12b, 13t, 13b, 18bl, 19t, 20bl; **Alcan International** 1, 16b, 43t, 44bl, 44/45b, 45tr; courtesy of **Chubb Security Group** 39tr; courtesy of **Rolls-Royce plc** BACK COVER, 20/21t, 23tr; by permission of **NASA** 30bl and **ZEFA** 41b, 42/43b.

Front cover: Aluminium can be extruded in one piece to make thin forms that can be torn easily, such as drinks cans. The ruby in the background is aluminium oxide.
Title page: Molten aluminium being poured from a crucible into a holding furnace.

This product is manufactured from sustainable managed forests. For every tree cut down at least one more is planted.

The demonstrations described or illustrated in this book are not for replication. The Publisher cannot accept any responsibility for any accidents or injuries that may result from conducting the experiments described or illustrated in this book.

Contents

Introduction

An element is a substance that cannot be broken down into a simpler substance by any known means. Each of the 92 naturally occurring elements is therefore one of the fundamental materials from which everything in the Universe is made. This book is about aluminium.

Aluminium

Aluminium, the third most common element on Earth after oxygen and silicon, and by far the most abundant metal on Earth, is now used very widely for everything from soft drink cans to car bodies to window frames. Compounds containing aluminium are found in materials as different as antacid medicines, the insulation materials in our homes and in the small white flecks (called vermiculite) in garden composts.

Aluminium is one of a number of soft and weak metals (like copper and tin) that scientists call "poor" metals. But aluminium alloys, mixtures of aluminium and other metals, produce materials as tough as steel.

The name aluminium comes from the word *alumen*, which is the Latin name for alum. Alum is an age-old material called a mordant, used for making dyes stick to fabrics.

Although it is so widely used today, aluminium has only recently come into use. This is because aluminium is so strongly attracted to oxygen that it can only be refined using huge amounts of electrical energy and electricity did not become readily available until this century.

Thus, it is sometimes known as the metal of the 20th century, just as iron was the metal of the 19th century.

Although electricity is relatively more plentiful and less expensive than it used to be, refining aluminium from its ore is still a costly process. This is why aluminium is often recycled. This way we do not have to "waste" energy refining more of the metal than we need to.

◄ For centuries aluminium could not easily be refined. This made it more precious than gold or silver! For much of the 19th century some important and wealthy people even used plates of aluminium in preference to fine china or silver.

For the same reason aluminium was used in settings that contained precious stones. The object shown here belonged to Napoleon III of France (b.1808 –d.1873).

► Aluminium can now be isolated from its ores relatively cheaply and has many everyday applications.

Minerals containing aluminium

Aluminium is one of the most common elements in the rocks of the Earth's surface, yet it is very rarely seen as the shiny metal we are used to seeing as soft drink cans etc. This is because aluminium easily combines with silicon and oxygen to make clay, the stuff of soils.

Micas

Micas are minerals made of sheets of aluminium, silicon and oxygen (known as aluminosilicates). Stacks of these sheets are connected together by metal ions; potassium and magnesium are among the most common.

The way the sheets are connected together is very important because it gives the minerals many of their properties. Sheet silicates all break up into thin flakes. For this reason, several of the sheet silicates are used as lubricants; for example, talcum powder is made from the mineral talc.

Mica is easily recognised as a silicate because it peels away into thin, almost transparent sheets. The main varieties are biotite, which is black, and muscovite, which is brown.

▶ Muscovite, a brown form of mica.

▶ Feldspar crystals are opaque and either pink or white. You can see them clearly in this granite sample.

Feldspar

Feldspars are common silicate minerals and found in most igneous rocks. To form an aluminium silicate of this kind, some of the oxygen atoms found in silica are replaced by aluminium atoms together with a small proportion of atoms of the metals potassium, sodium or calcium. Variety in the colour of feldspars is influenced by the proportions of these metals. For example, potassium feldspar is pale pink, whereas calcium feldspar is white.

Corundum

Corundum is a very hard, brown to black mineral, next only to diamond in its hardness. It is an aluminium oxide and is found next to where granite and other volcanic rocks occur. The material we call emery is made from corundum. Emery cloth is a common abrasive cloth, while emery powder is glued on to discs to make cutting and sanding tools.

▶ Aluminium sulphate (commonly known as alum).

clay: a microscopically small plate-like mineral that makes up the bulk of many soils. It has a sticky feel when wet.

crystal: a substance that has grown freely so that it can develop external faces. Compare with crystalline, where the atoms are not free to form individual crystals and amorphous where the atoms are arranged irregularly.

granite: an igneous rock with a high proportion of silica (usually over 65%). It has well-developed large crystals. The largest pink, grey or white crystals are feldspar.

igneous rock: a rock that has solidified from molten rock, either volcanic lava on the Earth's surface or magma deep underground. In either case the rock develops a network of interlocking crystals.

Kaolinite

Kaolinite is a clay mineral found in many of the world's soils. It is composed of sheets containing aluminium silicates. Kaolinite is a soft mineral whose crystals are too small to be seen even with an ordinary microscope.

Water molecules can be absorbed between the silicate sheets, which explains why soils shrink and swell with wetting and drying.

Kaolinite is mined as china clay and used for porcelain, pottery, the filler in medicinal tablets, and also to make smooth-textured paper.

Also...

The chemical formula for corundum is Al_2O_3 (aluminium oxide). There are three main types of feldspar, varying only in the metal that bonds the sheets together. The chemical formula for these feldspars is $KAlSi_3O_8$ (potassium feldspar), $CaAlSi_3O_8$ (calcium feldspar) and $NaAlSi_3O_8$ (sodium feldspar). The chemical formula for muscovite mica is $KAl_2(Si_3Al)O_{10}(OH)_2$ (potassium aluminosilicate).

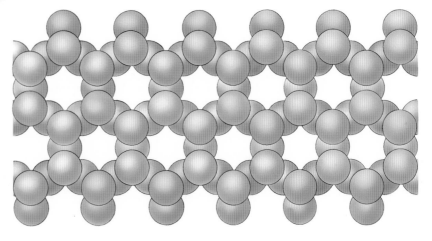

▼ The structure of many sheet minerals containing aluminium. Mica and kaolinite are good examples.

Gemstones containing aluminium

Whereas clay is made of tiny, unremarkable crystal plates that are too small to see without a powerful microscope, just occasionally, usually near where volcanoes have been active, aluminium combines with other elements to form some of the world's most remarkable crystals. These are gemstones such as sapphire and ruby.

Ruby

Ruby is a deep red crystal, one of the most prized of all gemstones. It is made mainly of aluminium and oxygen (aluminium oxide). This mineral, known as corundum, is transparent. But when it occurs with small amounts of another element, chromium, the colour changes to somewhere between pale rose and deep red.

The world's best rubies come from Myanmar (Burma), where they have been naturally weathered from rocks and washed by rivers to accumulate among river gravels.

Small rubies can now be made artificially, and these are routinely used in many of the world's lasers.

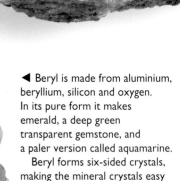

◄ Beryl is made from aluminium, beryllium, silicon and oxygen. In its pure form it makes emerald, a deep green transparent gemstone, and a paler version called aquamarine.

Beryl forms six-sided crystals, making the mineral crystals easy to identify. It is often found associated with volcanic rocks.

Emeralds are found where volcanic activity has altered limestones, often in the form of very large crystals. One of the largest crystals found weighed 200 tonnes! Needless to say, this was not of the transparent gemstone variety.

▼ Ruby set in the greenstone rock (zoisite) in which it is normally found. The hexagonal crystal system to which ruby belongs shows clearly in this example.

Ruby

bond: chemical bonding is either a transfer or sharing of electrons by two or more atoms. There are a number of types of chemical bond, some very strong (such as covalent bonds), others weak (such as hydrogen bonds). Chemical bonds form because the linked molecule is more stable than the unlinked atoms from which it formed. For example, the hydrogen molecule (H_2) is more stable than single atoms of hydrogen, which is why hydrogen gas is always found as molecules of two hydrogen atoms.

gemstone: a wide range of minerals valued by people, both as crystals (such as emerald) and as decorative stones (such as agate). There is no single chemical formula for a gemstone.

molecule: a group of two or more atoms held together by chemical bonds.

▼▶▶ Sapphire is a pale blue to deep violet form of corundum (see also page 6). The colours are produced by various amounts of the elements iron and titanium.
Most transparent corundum is known as sapphire, except for the red varieties which are called ruby (see page 8). Consequently, rubies and sapphires are often collected from the same deposits as the corundum that is used in industry. Sapphires can be made artificially and are used where hard wearing is vital, such as for the jewels of watch bearings. Sapphire can also be used as an abrasive.

Bauxite

Aluminium is found in every handful of soil you hold. It begins as a part of minerals in rocks such as granite. One of the most common minerals containing aluminium is pink or grey feldspar. To break down the feldspar and make clay, all that nature requires is water, warm weather and a very long period of time.

As rainwater washes over the surface of the feldspar, invisible chemical reactions occur. They are the same reactions that cause limestone statues on buildings to become weathered. The water, often with carbon dioxide gas from the air, rots the feldspar. The result is to release the elements in the feldspar and put them into solution. The aluminium quickly combines with silicon and oxygen to form clay. Because the clay is formed in water, it can no longer be destroyed by future rainfall. This is the secret to how aluminium is locked up in the world's clays and why it is so difficult to recover it.

The ore containing aluminium, called bauxite, forms in places where the aluminium compounds become especially concentrated.

◀▲ Open cast mining of bauxite in Jamaica. The waste rock and soil, called the overburden, is first removed and then the soft bauxite is dug out in large slices before being carried away using dumper-trucks. The overburden is used to fill in the areas already mined of its bauxite. The land can then be reclaimed.

feldspar: a mineral consisting of sheets of aluminium silicate. This is the mineral from which the clay in soils is made.

granite: an igneous rock with a high proportion of silica (usually over 65%). It has well-developed large crystals. The largest pink, grey or white crystals are feldspar.

ore: a rock containing enough of a useful substance to make mining it worthwhile.

Bauxite

The name bauxite comes from an ancient mine site at Les Baux in the south of France. Bauxite is a tropical soil material, full of clay like any other soil. Under hot, moist conditions some of the aluminium does not get locked up as clay, but instead forms sheets of aluminium oxide.

Aluminium oxide is usually colourless or greyish-white, and often forms alongside oxides of iron under humid tropical conditions. Iron oxides are red, and their bright colour masks the pale aluminium oxide. Most bauxite is found inside the tropical red (iron-containing) subsoil material known as laterite.

To get aluminium metal from bauxite the aluminium must be separated from both the oxygen and the oxides of iron.

Modern bauxite mines are located where the ores contain at least half their volume of aluminium oxide. All of the ores are soil layers and thus are always mined in shallow open-cast pits. Today the majority of bauxite comes from Guinea, Australia, Jamaica and Brazil.

The known reserves of bauxite will last for several hundred years if the consumption can be kept to present levels by careful recycling.

◀ Bauxite rock has an orange–red colour because of staining by iron oxides.

The process of manufacturing aluminium metal from bauxite has many stages. These are described here and on the following pages.

The aluminium industry begins

The history of the aluminium industry is quite short. The industrial method for separating aluminium from bauxite ore was only discovered in 1854, and the first aluminium was produced in 1859. But it is now one of the most important metal industries in the world.

How aluminium came to be refined

The aluminium industry had to wait for the development of electricity. In fact, the person who first separated aluminium from its ore was Danish professor Hans Christian Oersted, one of the pioneers of electricity.

However, only after discoveries in 1886 by Charles Martin Hall of Ohio, USA, and Paul L. T. Héroult of France, and in 1888 by Karl Joseph Bayer of Germany, did it become possible to refine large amounts of aluminium. Even then, large-scale processing did not get under way until the early part of the 20th century. This is because relatively cheap electricity supplies were needed and it took some time for the power generating industry to build generators large enough for the needs of an aluminium refinery. So, while iron was the metal of the 19th century, aluminium became the metal of the 20th century.

◄▲ In 1886 Charles Martin Hall (above) of the United States and Paul L. T. Héroult (below left) of France, discovered a way to dissolve alumina in molten cryolite and so make aluminium commercially. For this reason the process is called the Hall–Héroult process, and it is still in use today.

refining: separating a mixture into the simpler substances of which it is made. In the case of a rock, it means the extraction of the metal that is mixed up in the rock. In the case of oil it means separating out the fractions of which it is made.

◄ In this aluminium smelter (in Lochaber, Scotland) the power is provided by an on-site hydroelectric power plant. You can see the pipes that carry water to the power station turbines behind the main works. It takes about 20 kilowatt-hours of electricity to produce a kilogram of aluminium.

▼ Because of the low percentage of aluminium in the bauxite ore from which it is extracted, large machines with bucket scoops have to be used to gain economies of scale.

This bauxite from Ghana is awaiting processing into alumina.

Dissolving aluminium compounds

Aluminium is a reactive metal. When exposed to the air it immediately develops an oxide coating that prevents further corrosion. However, along with a few other metals, aluminium compounds can be dissolved by both acids and alkalis. Metal compounds with this special property are called amphoteric metals. This property has been exploited in the aluminium industry as a way of dissolving the aluminium compounds from bauxite while leaving the rest of the ore as a solid. This process, called the Bayer process, is shown on pages 16 and 17.

❶▲ The completely intact aluminium container used in the demonstration.

❷▶ Sodium hydroxide is poured into the container. It immediately begins to fizz and bubble.

amphoteric: a metal that will react with both acids and alkalis.

bauxite: an ore of aluminium, of which about half is aluminium oxide.

dissolve: to break down a substance in a solution without a resultant reaction.

solution: a mixture of a liquid and at least one other substance. Mixtures can be separated out by physical means, for example by evaporation and cooling.

❸◀ Underneath the bubbling a chemical reaction is taking place that is putting the aluminium into solution.

❹▼ The aluminium container after reaction. Notice that the container no longer has a bottom! The reaction time from start to finish was five minutes.

Also:

If sodium hydroxide (caustic soda – a powerful base), is poured into an aluminium saucepan a chemical reaction will occur that dissolves the aluminium. Cleaning aluminium cookware with caustic soda is therefore not to be recommended!

EQUATION: Dissolving aluminium in sodium hydroxide

Aluminium + sodium hydroxide+ water ⇨ sodium aluminate + hydrogen

$$2Al(s) \ + \ 2NaOH(aq) \ + \ 6H_2O(l) \ ⇨ \ 2NaAl(OH)_4(aq) \ + \ 3H_2(g)$$

Concentrating bauxite ore

Bauxite is a red rock-like material. It consists of aluminium oxide and a wide range of unwanted substances. To produce aluminium metal, the ore first has to be concentrated, thus removing the bulk of the impurities. Then it goes to a refinery, where the pure metal is produced. The concentrating stage is called the Bayer process.

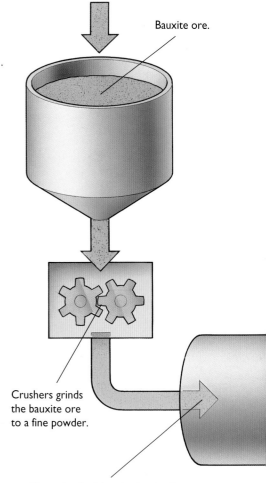

Bauxite ore.

Crushers grinds the bauxite ore to a fine powder.

The pulverised bauxite is mixed with sodium hydroxide (NaOH) at high pressure and temperature.

The Bayer process

Chemical reactions work most efficiently if the reactants have a large surface area. In the case of an ore, chemical reactions are made faster by pulverising the ore into a powder. Bauxite powder is mixed with crushed sodium hydroxide (soda ash) and calcium oxide (lime) and then mixed with water.

The chemical reaction takes place at high temperatures and pressures, making the aluminium oxide soluble (as sodium aluminate), so that it can be drained off into settling tanks. Here, any solids, such as pieces of sand and iron oxide, settle out. The alumina is a light brown liquid.

This liquid is drawn off and pumped into tall vats where it is allowed to cool. Inside the vats the liquid is stirred and tiny crystals of alumina form. The stirring causes the crystals to stick to each other until they are about the size of grains of sugar, then they sink to the bottom. The crystals are taken away and washed clear of any remaining liquid. They are then heat-dried, and the resulting alumina is a grey powder.

EQUATION: Production of alumina

(i) dissolving bauxite in sodium hydroxide

Bauxite + sodium hydroxide + water ⇨ sodium aluminate

$Al_2O_3(ore) + 2NaOH(aq) + 3H_2O$ ⇨ $2NaAl(OH)_4(aq)$ + impurities
red mud

(ii) recovering the sodium hydroxide to leave alumina

Sodium aluminate ⇨ alumina + sodium hydroxide

$2NaAl(OH)_4(aq)$ ⇨ $Al_2O_3 \cdot 3H_2O(s)$ + $2NaOH(aq)$

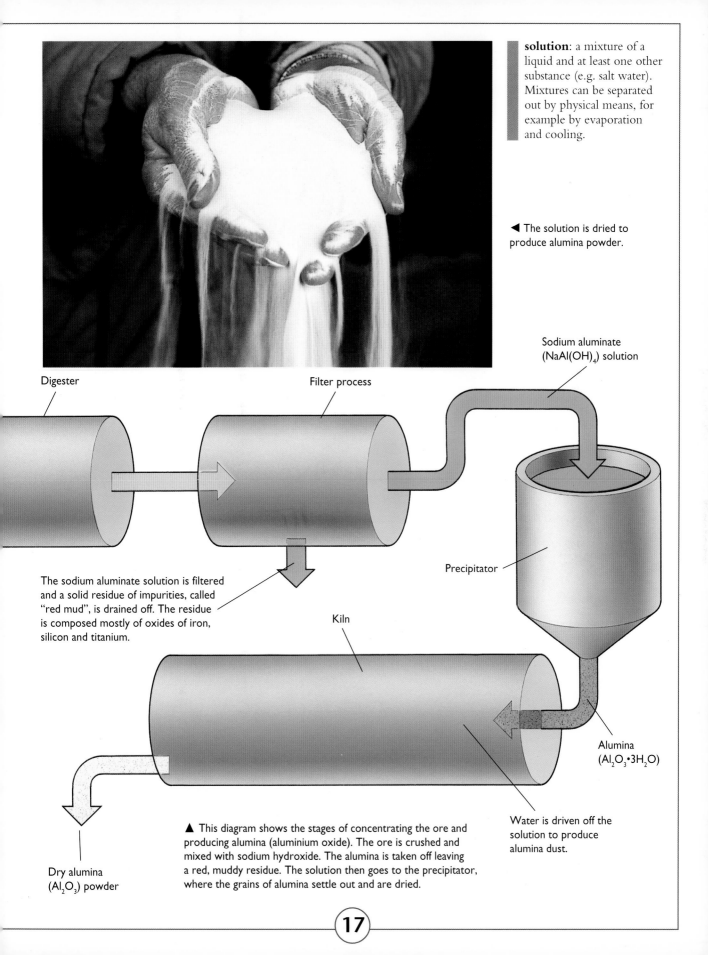

solution: a mixture of a liquid and at least one other substance (e.g. salt water). Mixtures can be separated out by physical means, for example by evaporation and cooling.

◄ The solution is dried to produce alumina powder.

Digester

Filter process

Sodium aluminate $(NaAl(OH)_4)$ solution

The sodium aluminate solution is filtered and a solid residue of impurities, called "red mud", is drained off. The residue is composed mostly of oxides of iron, silicon and titanium.

Precipitator

Kiln

Alumina $(Al_2O_3 \cdot 3H_2O)$

Water is driven off the solution to produce alumina dust.

▲ This diagram shows the stages of concentrating the ore and producing alumina (aluminium oxide). The ore is crushed and mixed with sodium hydroxide. The alumina is taken off leaving a red, muddy residue. The solution then goes to the precipitator, where the grains of alumina settle out and are dried.

Dry alumina (Al_2O_3) powder

Aluminium refining

The partly purified ore of bauxite, called alumina, is still a compound of aluminium and oxygen. To refine this to aluminium, the alumina has to be dissolved and the aluminium recovered by electrical means.

The process of using electricity to separate a metal from its rock ore is called electrolysis and takes place inside electrolytic cells. The alumina has to be liquified so that the aluminium compound will dissociate (break apart) into electrically charged particles called ions. Aluminium ions have a positive charge and can move through a solution to gather at the negatively charged electrode of the cell (the cathode).

Each cell uses a mere four to six volts, about the same as a dry cell used for a flashlight. But the current that flows is sometimes as much as 150,000 amps (equivalent to the maximum consumption of 300 households with all their appliances in use).

An aluminium smelter

Processing alumina happens on a large scale, but the electrical process cannot be done in a single large vat. Instead hundreds or even thousands of cells are used, made of steel with a carbon lining.

The process turns aluminium oxide into aluminium by removing the oxygen. This is called reduction.

The cells are first filled with a material called cryolite, which is heated to 980°C. Alumina will melt at a lower temperature when mixed with cryolite than if it were melted on its own. This saves electricity.

The alumina and cryolite mixture is then poured into the cell and rows of carbon electrodes are dipped into it. A current flows from the hanging electrodes to the carbon lining of the cell. At the bottom of the cells, embedded in the carbon lining, are collector plates (cathodes).

The electrical energy separates the aluminium ions from the oxygen ions and the aluminium collects on the plates at the bottom of the cell.

This process operates continuously, the molten aluminium being siphoned out of the cells and new alumina added from above. The aluminium can then be fed to mills and rolled into sheets, poured into moulds where it cools to make ingots for later use, or mixed with other metals to make alloys.

◄ This is the "cell room" of an aluminium smelting plant. Each cell is up to ten by four metres. Each cell makes about nine hundred kilograms of aluminium each day

▲ Aluminium is being poured into moulds to make ingots.

electrolysis: an electrical–chemical process that uses an electric current to cause the break up of a compound and the movement of metal ions in a solution. The process happens in many natural situations (as for example in rusting) and is also commonly used in industry for purifying (refining) metals or for plating metal objects with a fine, even metal coating.

ion: an atom, or group of atoms, that has gained or lost one or more electrons and so developed an electrical charge. Ions behave differently from electrically neutral atoms and molecules. They can move in an electric field, and they can also bind strongly to solvent molecules such as water. Positively charged ions are called cations; negatively charged ions are called anions. Ions carry electrical current through solutions.

reduction: the removal of oxygen from a substance. See also: oxidation.

EQUATION: Overall equations for the reduction of alumina to aluminium

Alumina ⇨ aluminium + oxygen

$$2Al_2O_3(s) \quad ⇨ \quad 4Al(s) \quad + \quad 3O_2(g)$$

Carbon + oxygen ⇨ carbon dioxide

$$3C(s) \quad + \quad 3O_2(g) \quad ⇨ \quad 3CO_2(g)$$

▼ The alumina smelter where an electrolysis process separates the aluminium from the oxide electrically, using the Hall–Héroult process.

Alumina (Al_2O_3) is dissolved in an electrolyte bath of cryolite (Na_3AlF_6) and aluminium fluoride (AlF_3).

Electrolytic cell or "pot".

Electrical current is passed through a graphite (carbon) anode

Siphon

Crucible

Other elements are added to create alloys.

Holding furnace

Molten aluminium

Ingot mould

Molten electrolyte

Electrical current is passed through the cathode

The properties of aluminium

Aluminium is soft and easy to shape but will become harder if it is cold-worked by, for example, hammering or pressing. This is called annealing.

The other important property of aluminium is that it can be welded, using a very high temperature source such as an arc-welding machine and a stick of aluminium. Aluminium can also be brazed (high temperature soldering using a rod of brass) or soldered with other metals (although the soldered joints suffer corrosion, making them unsuited for some uses). Aluminium can also be made harder when formed as an alloy.

◄ Aluminium is being extruded (pushed through a press containing a specially shaped die) to create the intricate shapes needed for the building industry. This is cladding.

▼ Aluminium alloys are widely used in aircraft parts, from the fuselage to the engines. Light weight and resistance to corrosion are important factors.

brazing: a form of soldering, in which brass is used as the joining metal.

soldering: joining together two pieces of metal using solder, an alloy with a low melting point.

welding: fusing two pieces of metal together using heat.

▼ Aluminium foil, thin sheets that are readily used for packaging.

◀ Lightweight coin from China.

Saving weight and fuel with aluminium

Aluminium is one of the lightest metals. This means that it will take less energy to move a piece of aluminium than, for example, a piece of steel of the same size. The saving in weight, and so in energy, is especially important in transport. If a steel container can be replaced by an aluminium alloy container, less fuel is needed to carry it, making the container cheaper to move.

The same is true of all the components of a vehicle. The engine block, the drive shafts, the radiator, the wheels and the body panels can all be made of aluminium alloys. The car thus weighs less and the fuel consumption improves. Unfortunately, the price also increases, which is why many vehicles are still made of steel.

Aluminium alloys

Pure aluminium is easy to bend and form into shape, but it is not very strong. An alloy is a mixture of various elements, designed to give the alloy special characteristics that pure metals do not have.

Common alloys

Normal refining processes do not remove all impurities from aluminium, so most commonly used industrial aluminium already has small amounts of iron, silicon and copper alloyed with it. Fortunately, alloying elements make aluminium stronger, but it remains easy to bend into shape.

The most common alloy of aluminium contains copper, magnesium, manganese, chromium, silicon, iron, nickel and zinc. Here silicon makes the aluminium flow better into casts of complicated shapes. Copper, on the other hand, makes the aluminium stronger. Adding 1.25% manganese makes the alloy exceptionally resistant to corrosion.

In all cases, the proportion of added elements is less than 10% of the final alloy; and because silicon mixes quite poorly in an alloy, no more than about 1.5% can be used.

Modern aircraft engine makers and power station turbine makers are working with new alloys that contain titanium or nickel to make lightweight aluminium even tougher.

Alloys that are forced into new shapes

For many purposes the aluminium alloy needs to be forced into a new shape, for example made into a tube or pressed into a car body panel. The best alloys for this have a slightly different composition to those used for casting. These alloys get stronger and harder as they are worked, often at high temperatures.

▲ Not all alloys make strong materials. Mercury, for example, forms an amalgam with aluminium on contact, forming a liquid metal. The holes in this sample of aluminium show where the mercury has "eaten" its way through the sheet.

▲ Aluminium alloys are commonly used for racing bicycles to give strength and light weight.

▲ Aluminium is often found in electric motors. It is also widely used for the pistons in engines.

▼ Wind turbines, such as those shown below, are often made with aluminium alloys so that the blade weight is kept to a minimum. In exposed locations, resistance to corrosion is also vital.

Special alloys

Aluminium alloys can be made as strong as steel. To make duralumin, which is both hard and strong, an alloy containing 4.5% copper, 0.6% manganese, and 1.5% magnesium is heated to about 495°C and then dipped in cold water. It is then left to "age" over the next few days, during which time it beomes extremely hard.

To make alclad, an alloy that has a soft aluminium centre and a hard, corrosion-resistant skin, an alloy of magnesium, zinc, copper, and chromium is heat-treated, quenched and allowed to "age". It is an ideal material for use on aircraft fuselages.

Where the aluminium alloy needs to retain strength at very high temperatures, nickel is alloyed with aluminium.

Making use of the reactivity of aluminium

Aluminium is the most reactive metal in common use. All metals more reactive than aluminium (calcium, sodium, etc.) are unstable and need special handling.

The reactivity of aluminium has advantages and disadvantages. One advantage is that it readily reacts with oxygen from the air, forming a gastight and invisible oxide layer on its surface that protects the metal from environmental corrosion.

On the other hand, being so reactive, the metal is very difficult to separate out from its ore and the costs of manufacture are high.

Aluminium and the reactivity series

Each metal reacts with the environment differently. Some, like potassium, are highly reactive; others, like gold, are very stable.

When two different metals are placed in a conducting solution (an electrolyte such as salt water), a natural battery is formed. In a battery, one of the electrodes (the anode) always corrodes, while the other (the cathode) becomes plated (or coated) with material from the corroding electrode.

Which electrode corrodes, and which is protected (becomes plated) depends on the positions of the metals in the reactivity series. Metals above those in the table become corroded; those lower down are protected.

Aluminium is more reactive than, for example, iron, so when aluminium and iron are placed together in salt water, for example, the aluminium will corrode rather than the iron (see opposite).

The reactivity table also helps to explain why aluminium is so difficult to extract from its ores compared with many other common metals. The higher up the table, the more energy it takes to separate the metal from its ores. For example, iron (in the middle of the table) can be smelted (a chemcial reaction involving heat and a reducing agent) whereas aluminium (near the top of the table) can only be refined using large amounts of electrical energy.

REACTIVITY SERIES	
Element	Reactivity
potassium	most reactive
sodium	
calcium	
magnesium	
aluminium	
manganese	
chromium	
zinc	
iron	
cadmium	
tin	
lead	
copper	
mercury	
silver	
gold	
platinum	least reactive

Aluminium as a sacrificial anode

Aluminium is bolted on to the keels of many ships to protect the main steel hull of the vessel.

When steel is placed in salt water it behaves like part of an electrical battery. In any battery there are two electrodes (a negative electrode or cathode, and a positive electrode or anode). As the battery works, one of the electrodes (the positive electrode, or anode) is corroded (used up).

There is a danger that the steel of a ship's hull will behave as the electrode that is used up and this can cause severe corrosion to the ship. However, if pieces of aluminium are placed on the hull below the water line, the aluminium is sacrificed instead, so protecting the hull.

Scientists call this effect "cathodic protection" because the hull of the ship is the cathode of the natural battery. It is far easier to replace chunks of aluminium bolted to the steel hull than to have to replace the whole hull! The secret to how this works lies in the fact that aluminium is much more reactive than iron. Other common cathodic protectors are zinc and magnesium.

cladding: a surface sheet of material designed to protect other materials from corrosion.

corrosion: the *slow* decay of a substance resulting from contact with gases and liquids in the environment. The term is often applied to metals. Rust is the corrosion of iron and steel.

oxide: a compound that includes oxygen and one other element.

rust: the corrosion of iron and steel.

▼ The hulls of ships are usually fitted with sacrificial anodes. As they are below the water line they are normally obscured except when the ship is in dry dock.

Anodising aluminium

The word anodising refers to a shiny protective coating that is often applied to aluminium to improve its looks. It is the opposite to the process of electroplating.

An anode is the electrode of a cell on which a substance oxidises (corrodes). Anodising is a process that uses electricity to produce this layer of oxide in a controlled way, in order to protect a surface from further corrosion.

The anodising process

Aluminium is placed in a chemical bath that will carry electricity, usually sulphuric or chromic acid. The aluminium acts as an anode. A cathode, often a carbon rod, is also placed in the bath. This makes an electrolytic cell. An electric current is passed through the cell and the surface of the aluminium immediately begins to change. The acid liberates oxygen at the anode and this combines with the aluminium to make aluminium oxide.

In air aluminium would develop a very thin surface coating of oxide, but in anodised aluminium, the oxide is a far thicker protective coating. It is also possible to add colour during this process. And because anodising is a change to the surface of the metal, it does not chip or wear off as a paint or plastic coating might.

Anodised aluminium is used as a decorative form of aluminium on hi-fi and other electrical equipment. It is used for vehicle parts, and can be used for both lighting and electrical fittings.

▼ These cowboy spurs are anodised so that they will not corrode even when used for long periods out in the rangelands.

anode: the negative terminal of a battery or the positive electrode of an electrolysis cell.

anodising: a process that uses the effect of electrolysis to make a surface corrosion-resistant.

electrolysis: an electrical–chemical process that uses an electric current to cause the break up of a compound and the movement of metal ions in a solution. The process happens in many natural situations (as for example in rusting) and is also commonly used in industry for purifying (refining) metals or for plating metal objects with a fine, even metal coating.

electroplating: depositing a thin layer of a metal onto the surface of another substance using electrolysis.

oxide: a compound that includes oxygen and one other element.

▲ There are two ways of providing a protective and decorative coating to aluminium products. The cheaper way is to apply a surface paint, but this is liable to wear off through handling. The better solution is to anodise the aluminium so that it remains protected and good-looking throughout the lifetime of the product.

◀ This saucepan is made from anodised aluminium. The aluminium oxide surface has the same hardness as other aluminium oxide minerals, such as sapphire, and so it is harder than stainless steel and even more scratch-resistant. By anodising the surface, manufacturers can use aluminium (which is reknowned for its good heat transmission properties, but usually suffers from being too soft for prolonged use) for high-quality, long-lasting cookware.

Aluminium as a conductor

Aluminium is a good conductor of both heat and electricity. It has found widespread applications in the electricity supply industry and also in places where good heat conductivity is needed, such as in radiators and cookware.

An electrical conductor

Aluminium conducts electricity about two-thirds as well as copper. When made as a special alloy, or mixture of metals, aluminium is an even better conductor than copper on a weight for weight basis. This is because aluminium is only one-third as dense as copper. It is also very much cheaper to use than copper, especially where large cables are needed, such as those slung between pylons or the main cables buried beneath the street. Today more than nine out of every ten kilometres of large-diameter electrical cable are made from aluminium rather than traditional copper.

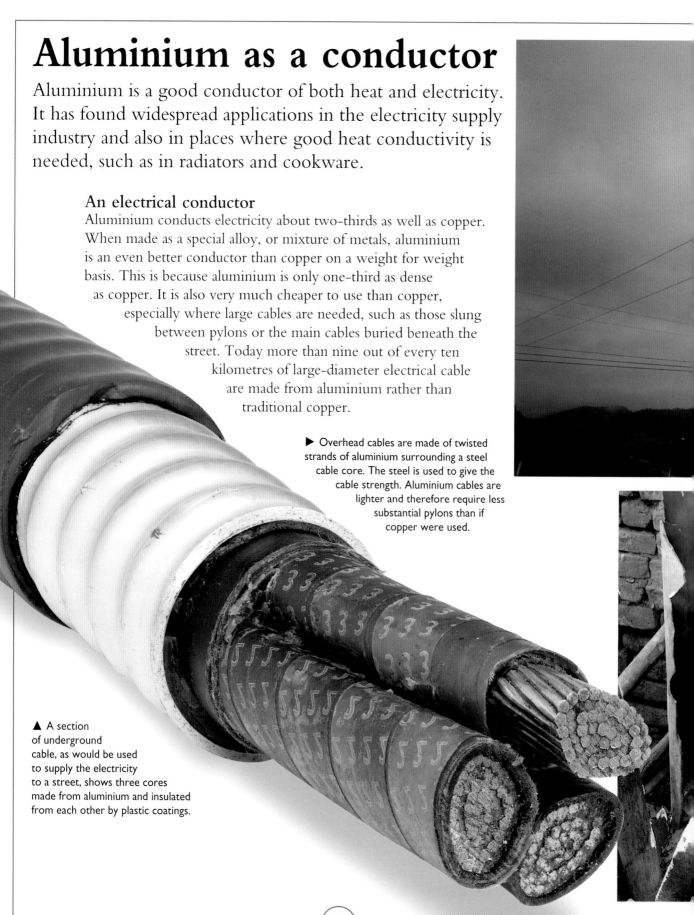

▶ Overhead cables are made of twisted strands of aluminium surrounding a steel cable core. The steel is used to give the cable strength. Aluminium cables are lighter and therefore require less substantial pylons than if copper were used.

▲ A section of underground cable, as would be used to supply the electricity to a street, shows three cores made from aluminium and insulated from each other by plastic coatings.

A heat conductor

Like all metals, aluminium conducts heat and so can be used either to carry heat away from a hot object or to bring heat to a cold object.

Aluminium is very much more expensive than steel, but where light weight is important, such as in aircraft, car or motorbike engines, aluminium is often the preferred choice. It is nearly twice as good a conductor, weight for weight, as copper, and about nine times as good as steel.

Because aluminium can be cast into detailed shapes and is easy to cut, it is sometimes used for the whole of the main engine part (known as the engine block). It is also used in the radiators of most engines. Aluminium used to conduct heat can easily be seen in motorbike engines, where the cooling fins show clearly.

◄ Aluminium is widely used for cooking utensils because of its good heat conductivity and light weight.

▲ Aluminium is used to carry heat away from sensitive electrical components on a circuit. The heating fins are light weight and so do not damage the board (as heavy steel might).

Aluminium as a reflector

One of the most useful features of aluminium is its excellent reflective properties.

Aluminium reflects about nine-tenths of heat reaching it. This means it can be used to reflect heat back inside a room or container to help keep it warm, or it can be used to reflect heat away from a container to keep it cool. For this reason aluminium is often used as part of the insulation of a house.

The reflective properties of aluminium can also be used in a quite different way. If an aircraft wants to jam the radar of an incoming missile, it can release large amounts of aluminium flakes or strips. These reflect the radar signals of the missile and makes it impossible to detect the correct target.

▼ Aluminium is increasingly used to conserve energy both in home heating and cooling and in the transportation industry. Aluminium storm doors and windows, insulation backed with aluminium foil, and aluminium siding are excellent insulators.

Aluminium and reflection in space

Aluminium is an excellent material to use in spacecraft and even in life support systems. This is because it is light, strong and has many useful heat reflecting properties.

The space environment is very harsh. People in space suits can be overheated while they are directly in sunlight, but they can begin to freeze when they are in the shade.

Aluminium helps to prevent the extremes of heat gain and loss by reflecting the radiated heat away from the space suit and also reflecting the body heat back in. This keeps down the amount of heating or cooling equipment that has to be fitted on to the space suit.

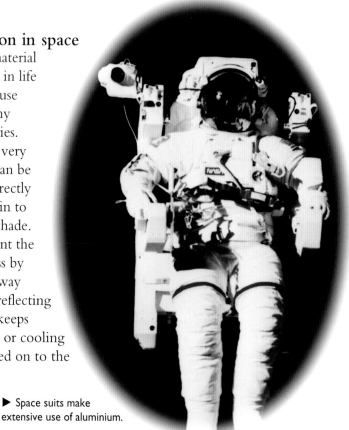

▶ Space suits make extensive use of aluminium.

Aluminium is a poor "radiator"

One of the special features of aluminium is that it is very poor at radiating heat. This means that, for example, if the sun shines on a piece of aluminium, any heat energy not reflected will cause the aluminium to heat up, but very little of the heat will be radiated.

Aluminium shares its heat by conduction or by heating the surrounding air, so that heat is lost by convection.

This means that a sheet of aluminium used to protect an object from the Sun will be much more effective than a sheet of steel. The aluminium will absorb the heat whereas the steel would absorb the heat on one side and radiate it from the other.

conduction: (i) the exchange of heat (heat conduction) by contact with another object or (ii) allowing the flow of electrons (electrical conduction).

convection: the exchange of heat energy with the surroundings produced by the flow of a fluid due to being heated or cooled.

radiation: the exchange of energy with the surroundings through the transmission of waves or particles of energy. Radiation is a form of energy transfer that can happen through space; no intervening medium is required (as would be the case for conduction and convection).

▼ Aluminium is often used instead of silver on large mirrors. It has good reflective properties, but is far less expensive than silver. The mirrors shown here are part of a solar generating plant.

Aluminium containers

The food and drink industry is the world's biggest user of aluminium. Factories that make aluminium containers account for about one-third of the world demand for aluminium.

By far the greatest demand is for aluminium cans, which have replaced the more traditional tin-plated steel can. The softness of aluminium allows the cans to be pushed into shape (extruded), while the force needed to tear the aluminium by the ring pull opener on drink cans is small because the aluminium has relatively little strength.

Modern packaging uses aluminium interleaved with plastic and paper to make other forms of carton which can be sealed. This allows liquids to be kept for long periods without the need for refrigeration.

On a quite different scale, because aluminium resists attack by some acids, it can be used to transport and store them.

Cooking utensils

One of the first uses for aluminium was for cooking utensils and many pans are still made from aluminium, either with or without nonstick coatings. High-quality utensils are hardened by anodising (see page 26).

The food processing industry uses aluminium for utensils such as steamers as well as for its packaging.

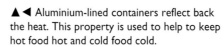

▲◀ Aluminium-lined containers reflect back the heat. This property is used to help to keep hot food hot and cold food cold.

Aluminium is also easy bent to shape while the aluminium oxide surface does not react with food or become corroded.

extrusion: forming a shape by pushing it through a die. For example, toothpaste is extruded through the cap (die) of the toothpaste tube.

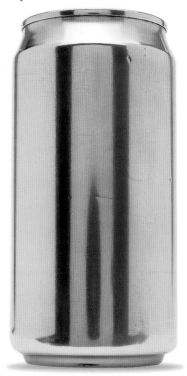

Contrasts in canning

A "tin can" is made from steel that has been coated with tin and sealed by soldering. The tin coating can be scratched away, allowing the steel to rust. Aluminium cans are lighter, do not rust, and can be extruded, thus removing the need for soldered seals. It is also easier to paint a decoration directly onto aluminium, so no more costly paper wrappers are used on steel cans.

▼▶ Aluminium can be made thin enough so that it can be torn. The special shapes on the top of many cans are designed to fail and produce a clean pouring opening. The lever-arrangement on the can is to save the can pulls from being discarded and polluting the environment.

Aluminium and acid containers

Aluminium has the advantage of being resistant to attack by a variety of chemicals. Although it is readily corroded by most alkalis (which attack the surface oxide film) aluminium does not react with ammonia and so it can be used to store and transport it.

Aluminium is far more resistant to neutral or acid solutions, in particular, acetic acid and concentrated nitric acid. For this reason many transport and storage containers for these chemicals are made from aluminium.

Aluminium oxide for separating mixtures

Aluminium oxide (alumina) is one of the most common yet overlooked of compounds. It occurs in all of the world's soils, combining with silicon compounds to make tiny particles of clay. Clays are important to a soil because they can hold nutrients (cations) on their surfaces, so in general a fertile soil is a soil with clay.

The way that aluminium oxide acts as a filter can be demonstrated by pouring a mixture of dyes in solution through a tube containing aluminium oxide powder.

❸◀ The column of aluminium oxide is prepared by filling it with acetone.

Acetone

Aluminium oxide

Cotton wool support

Rubber stopper with hole

❶◀ Most organic substances are made of a variety of compounds. Beetroot is a good example. It is first prepared by crushing it using a pestle and mortar.

❷▶ The beetroot juice is diluted with a liquid in which it can dissolve (in this case acetone).

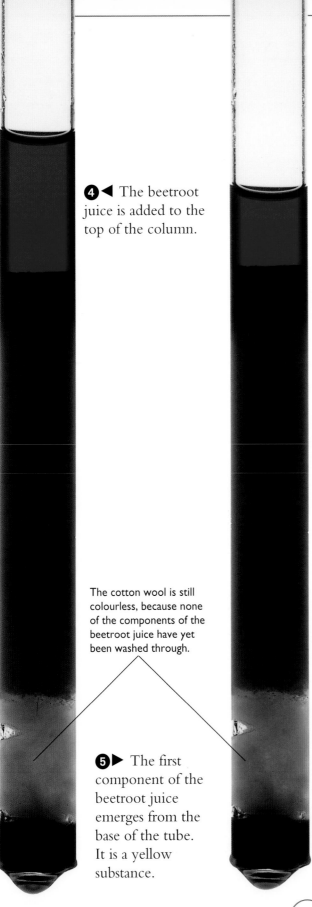

❹◀ The beetroot juice is added to the top of the column.

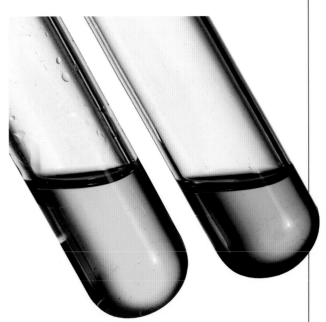

The cotton wool is still colourless, because none of the components of the beetroot juice have yet been washed through.

❺▶ The first component of the beetroot juice emerges from the base of the tube. It is a yellow substance.

❻▲ These are the first two components collected from the tube. Notice they are slightly different colours. The substance on the right is chlorophyll (the green pigment in plants), while that on the left is called xanthophyl.

Also...

The process demonstrated here is an example of chromatography, the use of a compound that does not react, to separate out the components of a complex substance (in this case a mixture of vegetable dyes). A substance like aluminium oxide, used in this way, is called a stationary phase. Substances easily attach to its surface (a process called adsorption) and are also easily washed off again. Each substance "sticks" to the aluminium oxide to a different degree, so that the least firmly stuck can be washed off most easily. The next one then washes off, and so on. As a result, the various compounds making up the original substance wash out of the base of the column one at a time, and can be collected separately.

Aluminium sulphate (alum)

Alum is a compound of aluminium sulphate. It is used as a mordant, that is a substance that will absorb dyes and so allow some natural fabrics to be coloured. The dye-absorbing property of alum is shown here.

For many thousands of years, people knew that clay-rich rocks contained a useful substance, even though they could not extract it properly or find out what it was. So they called it the metal of clay. Alum was eventually obtained by heating suitable rocks, often clay-rich rocks called shales, in sulphuric acid.

Today alum is still used as a mordant to help fix dyes in cotton fabric. The fabrics are first dipped in a solution containing alum and then in a solution of alkali so that a precipitate of aluminium hydroxide (the real mordant) is formed between the fibres. As the cotton fabric dries, the tiny particles of aluminium hydroxide remain in the fabric, and the dye sticks to them. Alum is also used as a filler in paper and to purify water.

◄ Freshly precipitated aluminium hydroxide, a nearly colourless gelatinous solid, shown here in a water suspension.

► A purple vegetable dye

EQUATION: Precipitating aluminium hydroxide mordant

Aluminium sulphate (alum) + ammonia solution + water ⇨ *aluminium hydroxide + ammonium sulphate*

$$Al_2(SO_4)_3(aq) \quad + \quad 6NH_3(aq) \quad + \quad 6H_2O(l) \quad \Rightarrow \quad 2Al(OH)_3(s) \quad + \quad 3(NH_4)_2SO_4(aq)$$

▲ A traditionally dyed cotton rug from Afghanistan.

dye: a coloured substance that will stick to another substance, so that both appear coloured.

gelatinous: a term meaning made with water. Because a gelatinous precipitate is mostly water, it is of a similar density to water and will float or lie suspended in the liquid.

mordant: any chemical that allows dyes to stick to other substances.

suspension: tiny particles suspended in a liquid.

Mordants for dyeing

Dyes will not easily stick to some natural fabrics such as cotton. So the fabric has to be treated with a special chemical that attracts the fabric and also the dye.

Alum is a mordant, a material that will mix with water in a solution to produce a precipitate that sticks fast to cotton fabric dipped in it.

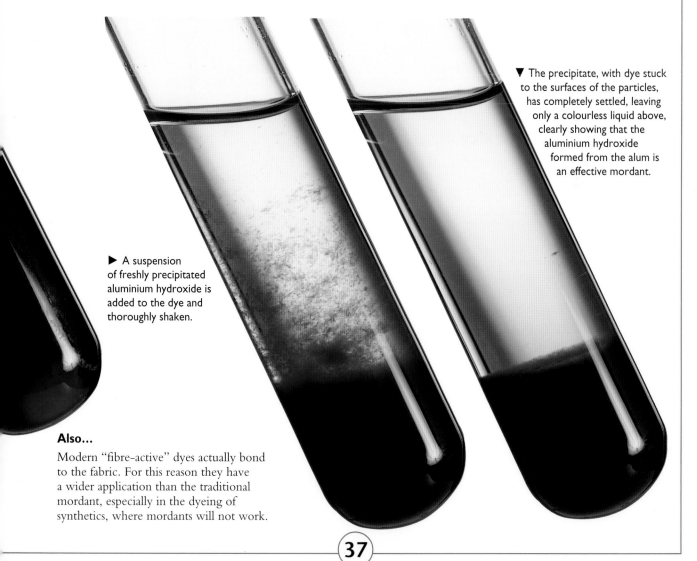

▶ A suspension of freshly precipitated aluminium hydroxide is added to the dye and thoroughly shaken.

▼ The precipitate, with dye stuck to the surfaces of the particles, has completely settled, leaving only a colourless liquid above, clearly showing that the aluminium hydroxide formed from the alum is an effective mordant.

Also...

Modern "fibre-active" dyes actually bond to the fabric. For this reason they have a wider application than the traditional mordant, especially in the dyeing of synthetics, where mordants will not work.

Aluminium compounds as foaming agents

Aluminium compounds can be used both as a source of carbon dioxide gas and a source of foam. Carbon dioxide gas can, in turn, be used to put out fires. For this reason such combinations were used to make liquid-type fire extinguishers for many years.

In the extinguisher, the two reagents (the liquids that will react) are kept apart until the extinguisher is to be used. Then, a knob on the extinguisher is struck, breaking the seal between the liquids and causing them to react.

The reaction produces a gelatinous precipitate of aluminium hydroxide and carbon dioxide gas. The gas cannot easily escape through this sticky liquid, and instead forms bubbles inside it. The result is a foam containing carbon dioxide that immediately squirts from the extinguisher nozzle. This has the effect of blanketing the fire with materials that will not burn, thus preventing oxygen from feeding the flames.

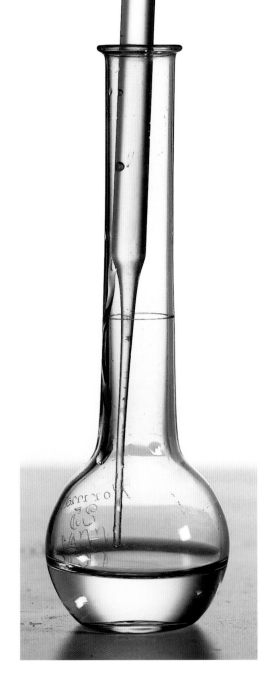

❶▶ Imagine this flask as the inside of a fire extinguisher. Two colourless liquids are kept separate and do not react. The lower one is concentrated sodium carbonate solution, the upper one concentrated aluminium sulphate solution.

EQUATION: Fire extinguishing

Aluminium sulphate + sodium carbonate + water ⇨ aluminium hydroxide + carbon dioxide + sodium sulphate

$$Al_2(SO_4)_3(aq) \ + \ 3Na_2CO_3(aq) \ + \ 3H_2O(l) \ \Rightarrow \ 2Al(OH)_3(s) \ + \ 3CO_2(g) \ + \ 3Na_2SO_4(aq)$$

foam: a substance that is sufficiently gelatinous to be able to contain bubbles of gas. The gas bulks up the substance, making it behave as though it were semi-rigid.

gelatinous: a term meaning made with water. Because a gelatinous precipitate is mostly water, it is of a similar density to water and will float or lie suspended in the liquid.

reagent: a starting material for a reaction.

▼ Carbon dioxide is used in fire extinguishers because it is not combustible. Surrounding a burning object with carbon dioxide therefore deprives the object of oxygen and the fire goes out. Carbon dioxide is also a non-polluting gas.

❷ ◄ Imagine that the fire extinguisher has been given a sharp tap to break the phial containing the aluminium sulphate solution, allowing it to mix with the sodium carbonate solution. The result is an immediate reaction, producing a gelatinous foam of carbon dioxide gas that froths up and does not disperse. These agents together make for an efficient fire extinguishing combination.

Why recycling is vital

Aluminium is essential to many kinds of manufacturing, but making it from bauxite ore requires a great deal of expensive energy. In countries that refine aluminium, one-hundredth of all the electricity made in power stations may be used to run the refineries. It takes about 14 kilowatt-hours to refine each kilogram of alumina. (About the same as 14 single-bar electric fires running for one hour).

▲ Recycled aluminium cast into blocks.

But once aluminium has been refined and used, it can be melted down and recycled using just one-twentieth of the energy it took to make it in the first place. By saving this amount of energy, not only can resources like coal and oil be saved, but less carbon dioxide (a greenhouse gas) and sulphur and nitrogen oxides (acid rain gases) are released by power stations. For all these reasons it makes sense to recycle.

Recycling now accounts for about one-half of the total use of aluminium.

▲ A few cans thrown in a waste basket may not seem very relevant to saving the world's aluminium resources…

acid rain: rain that is contaminated by acid gases such as sulphur dioxide and nitrogen oxides released by pollution.

Greenhouse Effect: an increase of the global air temperature as a result of heat released from burning fossil fuels being absorbed by carbon dioxide in the atmosphere.

refining: separating a mixture into the simpler substances of which it is made. In the case of a rock, it means the extraction of the metal that is mixed up in the rock.

▲ … until you see the result of countrywide collections.

Aluminium and the environment

Aluminium is a very reactive element and therefore is normally found locked up as compounds. Most compounds are not at all harmful in the environment and, as in the case of clay minerals, they are positively important.

The only time that aluminium becomes a problem in the environment is as an indirect result of various pollutants that people create.

Of these the most serious by far is acid rain, which increases the acidity of water in the soil. If soils do not contain a buffer to cope with this onslaught, the soil may well become acid enough for aluminium to go into solution and find its way into water supplies for animals and plants. As aluminium is a toxic substance, the effects can be serious.

The other important environmental effect of aluminium occurs during mining. Aluminium occurs in surface sheets, and so its recovery destroys large areas of land. Most mines are found in the tropics, and many in tropical rainforests, where conditions are not conducive to the recovery of the land once it has been disturbed. Furthermore, only a small amount of the ore is transported to refineries, the majority, known as red mud, is often allowed to leave the mines, where it can pollute nearby streams and coasts. There are severe red mud pollution problems in Jamaica, for example.

▶ Red mud from the chemical processing of bauxite can pollute rivers. In this picture from Jamaica it has been dammed in to form a dead lake.

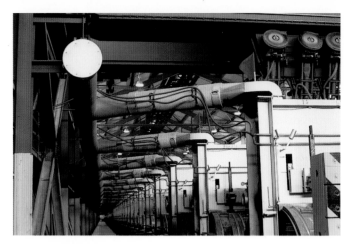

▲ Gases and dust are collected in special hoods attached to each electrolytic cell.

acidity: a general term for the strength of an acid in a solution.

buffer: a chemistry term meaning a mixture of substances in solution that resists a change in the acidity or alkalinity of the solution.

clay: a microscopically small plate-like mineral that makes up the bulk of many soils. It has a sticky feel when wet.

ion: an atom, or group of atoms, that has gained or lost one or more electrons and so developed an electrical charge.

pH: a measure of the hydrogen ion concentration in a liquid. Neutral is pH 7.0; numbers greater than this are alkaline, smaller numbers are acidic.

Aluminium and acid rain

Aluminium is insoluble if the pH of the soil is greater than 5. Only acid soils, such as podzols, which occur in areas of high rainfall or on acid parent materials, are therefore likely to cause aluminium to go into solution. So, under normal circumstances, aluminium compounds are not released into the soil or the water supply and are only taken up in small amounts by the body. More recently, the phenomenon of acid rain has changed this in some areas.

Acid rain occurs when polluting gases created by factories and vehicles, mix with water vapour in the air and then eventually fall as acidic raindrops or snow. This extra acid gets into the soil where it begins to act on the abundant aluminium in the clay crystals.

Acid rain releases aluminium compounds, which can then be taken up by plant roots where they can damage or even kill plants.

The aluminium compounds also flow into nearby rivers and lakes where they are absorbed by fish and other animals. Again the effects can be disastrous. In fact, many plants and animals die of aluminium poisoning. It is important to remember, though, that the real problem is not the presence of aluminium but the acid rain we produce by polluting the atmosphere with sulphur and nitrogen compounds.

Key facts about...

Aluminium

A soft, silvery-white metal, chemical symbol is Al

Can be toxic to plants if released into the soil by acid rain

Lightweight metal

Good conductor of electricity

Easily moulded to other shapes

Good conductor of heat

Melts at a lower temperature than most other metals (660°C)

Extremely strong when combined with other metals

Has no taste

Third most common element, making 8% of the Earth's crust

Atomic number 13, atomic weight about 27

SHELL DIAGRAMS

The shell diagram on this page represents an atom of the element aluminium. The total number of electrons is shown in the relevant orbitals, or shells, around the central nucleus.

Electron shell

Electron

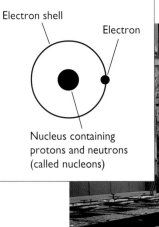

Nucleus containing protons and neutrons (called nucleons)

▶ ▼ Molten aluminium is siphoned from the electrolytic cells (left) into a holding furnace where it is kept at 690 to 745°C for two to three hours. Other elements can then be added to create the desired alloys. The resultant molten metal is then poured into moulds of the desired ingot shape (below), from which aluminium products are manufactured. Careful control of the cooling of the aluminium produces a good internal structure to the metal casts (bottom left).

The Periodic Table

The Periodic Table sets out the relationships among the elements of the Universe. According to the Periodic Table, certain elements fall into groups. The pattern of these groups has, in the past, allowed scientists to predict elements that had not at that time been discovered. It can still be used today to predict the properties of unfamiliar elements.

The Periodic Table was first described by a Russian teacher, Dmitry Ivanovich Mendeleev, between 1869 and 1870. He was interested in writing a chemistry textbook, and wanted to show his students that there were certain patterns in the elements that had been discovered. So he set out the elements (of which there were 57 at the time) according to their known properties. On the assumption that there was pattern to the elements, he left blank spaces where elements seemed to be missing. Using this first version of the Periodic Table, he was able to predict in detail the chemical and physical properties of elements that had not yet been discovered. Other scientists began to look for the missing elements, and they soon found them.

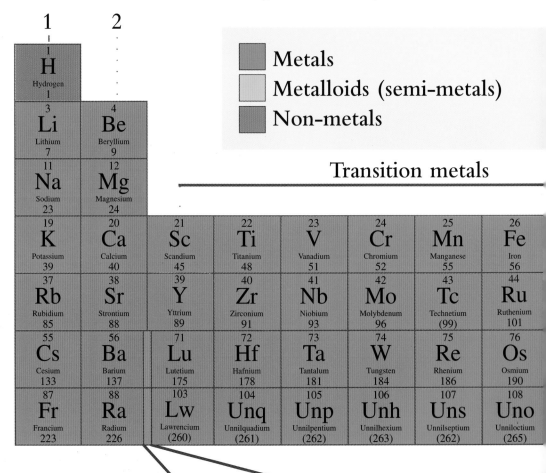

Hydrogen did not seem to fit into the table, so he placed it in a box on its own. Otherwise the elements were all placed horizontally. When an element was reached with properties similar to the first one in the top row, a second row was started. By following this rule, similarities among the elements can be found by reading up and down. By reading across the rows, the elements progressively increase their atomic number. This number indicates the number of positively charged particles (protons) in the nucleus of each atom. This is also the number of negatively charged particles (electrons) in the atom.

The chemical properties of an element depend on the number of electrons in the outermost shell.

Atoms can form compounds by sharing electrons in their outermost shells. This explains why atoms with a full set of electrons (like helium, an inert gas) are unreactive, whereas atoms with an incomplete electron shell (such as chlorine) are very reactive. Elements can also combine by the complete transfer of electrons from metals to non-metals and the compounds formed contain ions.

Radioactive elements lose particles from their nucleus and electrons from their surrounding shells. As a result their atomic number changes and they become new elements.

Atomic (proton) number — 13 — Symbol
Al
Aluminium — Name
27
Approximate relative atomic mass
(Approximate atomic weight)

3	4	5	6	7	0
					2 **He** Helium 4
5 **B** Boron 11	6 **C** Carbon 12	7 **N** Nitrogen 14	8 **O** Oxygen 16	9 **F** Fluorine 19	10 **Ne** Neon 20
13 **Al** Aluminium 27	14 **Si** Silicon 28	15 **P** Phosphorus 31	16 **S** Sulphur 32	17 **Cl** Chlorine 35	18 **Ar** Argon 40

27 **Co** Cobalt 59	28 **Ni** Nickel 59	29 **Cu** Copper 64	30 **Zn** Zinc 65	31 **Ga** Gallium 70	32 **Ge** Germanium 73	33 **As** Arsenic 75	34 **Se** Selenium 79	35 **Br** Bromine 80	36 **Kr** Krypton 84
45 **Rh** Rhodium 103	46 **Pd** Palladium 106	47 **Ag** Silver 108	48 **Cd** Cadmium 112	49 **In** Indium 115	50 **Sn** Tin 119	51 **Sb** Antimony 122	52 **Te** Tellurium 128	53 **I** Iodine 127	54 **Xe** Xenon 131
77 **Ir** Iridium 192	78 **Pt** Platinum 195	79 **Au** Gold 197	80 **Hg** Mercury 201	81 **Tl** Thallium 204	82 **Pb** Lead 207	83 **Bi** Bismuth 209	84 **Po** Polonium (209)	85 **At** Astatine (210)	86 **Rn** Radon (222)

109 **Une** Unnilennium (266)

61 **Pm** Promethium (145)	62 **Sm** Samarium 150	63 **Eu** Europium 152	64 **Gd** Gadolinium 157	65 **Tb** Terbium 159	66 **Dy** Dysprosium 163	67 **Ho** Holmium 165	68 **Er** Erbium 167	69 **Tm** Thulium 169	70 **Yb** Ytterbium 173
93 **Np** Neptunium (237)	94 **Pu** Plutonium (244)	95 **Am** Americium (243)	96 **Cm** Curium (247)	97 **Bk** Berkelium (247)	98 **Cf** Californium (251)	99 **Es** Einsteinium (252)	100 **Fm** Fermium (257)	101 **Md** Mendelevium (258)	102 **No** Nobelium (259)

Understanding equations

As you read through this book, you will notice that many pages contain equations using symbols. If you are not familiar with these symbols, read this page. Symbols make it easy for chemists to write out the reactions that are occurring in a way that allows a better understanding of the processes involved.

Symbols for the elements

The basis of the modern use of symbols for elements dates back to the 19th century. At this time a shorthand was developed using the first letter of the element wherever possible. Thus "O" stands for oxygen, "H" stands for hydrogen

and so on. However, if we were to use only the first letter, then there could be some confusion. For example, nitrogen and nickel would both use the symbols N. To overcome this problem, many elements are symbolised using the first two letters of their full name, and the second letter is lowercase. Thus although nitrogen is N, nickel becomes Ni. Not all symbols come from the English name; many use the Latin name instead. This is why, for example, gold is not G but Au (for the Latin *aurum*) and sodium has the symbol Na, from the Latin *natrium*.

Compounds of elements are made by combining letters. Thus the molecule carbon

Written and symbolic equations

In this book, important chemical equations are briefly stated in words (these are called word equations), and are then shown in their symbolic form along with the states.

What reaction the equation illustrates

EQUATION: The formation of calcium hydroxide

Word equation

Calcium oxide + water ⇨ calcium hydroxide

Symbol equation

$$CaO(s) \quad + \quad H_2O(l) \quad ⇨ \quad Ca(OH)_2(aq)$$

heated

Sometimes you will find additional descriptions below the symbolic equation.

Symbol showing the state: *s* is for solid, *l* is for liquid, *g* is for gas and *aq* is for aqueous.

Diagrams

Some of the equations are shown as graphic representations.

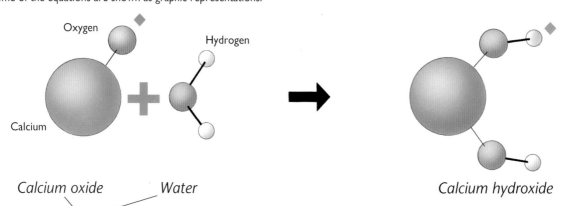

Oxygen

Hydrogen

Calcium

Calcium oxide   *Water*

Calcium hydroxide

Sometimes the written equation is broken up and put below the relevant stages in the graphic representation.

monoxide is CO. By using lowercase letters for the second letter of an element, it is possible to show that cobalt, symbol Co, is not the same as the molecule carbon monoxide, CO.

However, the letters can be made to do much more than this. In many molecules, atoms combine in unequal numbers. So, for example, carbon dioxide has one atom of carbon for every two of oxygen. This is shown by using the number 2 beside the oxygen, and the symbol becomes CO_2.

In practice, some groups of atoms combine as a unit with other substances. Thus, for example, calcium bicarbonate (one of the compounds used in some antacid pills) is written $Ca(HCO_3)_2$. This shows that the part of the substance inside the brackets reacts as a unit and the "2" outside the brackets shows the presence of two such units.

Some substances attract water molecules to themselves. To show this a dot is used. Thus the blue form of copper sulphate is written $CuSO_4.5H_2O$. In this case five molecules of water attract to one of copper sulphate.

When you see the dot, you know that this water can be driven off by heating; it is part of the crystal structure.

In a reaction substances change by rearranging the combinations of atoms. The way they change is shown by using the chemical symbols, placing those that will react (the starting materials, or reactants) on the left and the products of the reaction on the right. Between the two, chemists use an arrow to show which way the reaction is occurring.

It is possible to describe a reaction in words. This gives word equations, which are given throughout this book. However, it is easier to understand what is happening by using an equation containing symbols. These are also given in many places. They are not given when the equations are very complex.

In any equation both sides balance; that is, there must be an equal number of like atoms on both sides of the arrow. When you try to write down reactions, you, too, must balance your equation; you cannot have a few atoms left over at the end!

The symbols in brackets are abbreviations for the physical state of each substance taking part, so that (s) is used for solid, (l) for liquid, (g) for gas and (aq) for an aqueous solution, that is, a solution of a substance dissolved in water.

Atoms and ions
Each sphere represents a particle of an element. A particle can be an atom or an ion. Each atom or ion is associated with other atoms or ions through bonds – forces of attraction. The size of the particles and the nature of the bonds can be extremely important in determining the nature of the reaction or the properties of the compound.

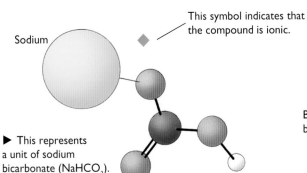

Sodium

This symbol indicates that the compound is ionic.

▶ This represents a unit of sodium bicarbonate ($NaHCO_3$).

The term "unit" is sometimes used to simplify the representation of a combination of ions.

Chemical symbols, equations and diagrams
The arrangement of any molecule or compound can be shown in one of the two ways shown below, depending on which gives the clearer picture. The left-hand diagram is called a ball-and-stick diagram because it uses rods and spheres to show the structure of the material. This example shows water, H_2O. There are two hydrogen atoms and one oxygen atom.

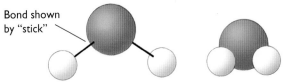

Bond shown by "stick"

Colours too
The colours of each of the particles help differentiate the elements involved. The diagram can then be matched to the written and symbolic equation given with the diagram. In the case above, oxygen is red and hydrogen is grey.

Glossary of technical terms

absorb: to soak up a substance. Compare to adsorb.

acetone: a petroleum-based solvent.

acid: compounds containing hydrogen which can attack and dissolve many substances. Acids are described as weak or strong, dilute or concentrated, mineral or organic.

acidity: a general term for the strength of an acid in a solution.

acid rain: rain that is contaminated by acid gases such as sulphur dioxide and nitrogen oxides released by pollution.

adsorb/adsorption: to "collect" gas molecules or other particles on to the *surface* of a substance. They are not chemically combined and can be removed. (The process is called "adsorption".) Compare to absorb.

alchemy: the traditional "art" of working with chemicals that prevailed through the Middle Ages. One of the main challenges of alchemy was to make gold from lead. Alchemy faded away as scientific chemistry was developed in the 17th century.

alkali: a base in solution.

alkaline: the opposite of acidic. Alkalis are bases that dissolve, and alkaline materials are called basic materials. Solutions of alkalis have a pH greater than 7.0 because they contain relatively few hydrogen ions.

alloy: a mixture of a metal and various other elements.

alpha particle: a stable combination of two protons and two neutrons, which is ejected from the nucleus of a radioactive atom as it decays. An alpha particle is also the nucleus of the atom of helium. If it captures two electrons it can become a neutral helium atom.

amalgam: a liquid alloy of mercury with another metal.

amino acid: amino acids are organic compounds that are the building blocks for the proteins in the body.

amorphous: a solid in which the atoms are not arranged regularly (i.e. "glassy"). Compare with crystalline.

amphoteric: a metal that will react with both acids and alkalis.

anhydrous: a substance from which water has been removed by heating. Many hydrated salts are crystalline. When they are heated and the water is driven off, the material changes to an anhydrous powder.

anion: a negatively charged atom or group of atoms.

anode: the negative terminal of a battery or the positive electrode of an electrolysis cell.

anodising: a process that uses the effect of electrolysis to make a surface corrosion-resistant.

antacid: a common name for any compound that reacts with stomach acid to neutralise it.

antioxidant: a substance that prevents oxidation of some other substance.

aqueous: a solid dissolved in water. Usually used as "aqueous solution".

atom: the smallest particle of an element.

atomic number: the number of electrons or the number of protons in an atom.

atomised: broken up into a very fine mist. The term is used in connection with sprays and engine fuel systems.

aurora: the "northern lights" and "southern lights" that show as coloured bands of light in the night sky at high latitudes. They are associated with the way cosmic rays interact with oxygen and nitrogen in the air.

basalt: an igneous rock with a low proportion of silica (usually below 55%). It has microscopically small crystals.

base: a compound that may be soapy to the touch and that can react with an acid in water to form a salt and water.

battery: a series of electrochemical cells.

bauxite: an ore of aluminium, of which about half is aluminium oxide.

becquerel: a unit of radiation equal to one nuclear disintegration per second.

beta particle: a form of radiation in which electrons are emitted from an atom as the nucleus breaks down.

bleach: a substance that removes stains from materials either by oxidising or reducing the staining compound.

boiling point: the temperature at which a liquid boils, changing from a liquid to a gas.

bond: chemical bonding is either a transfer or sharing of electrons by two or more atoms. There are a number of types of chemical bond, some very strong (such as covalent bonds), others weak (such as hydrogen bonds). Chemical bonds form because the linked molecule is more stable than the unlinked atoms from which it formed. For example, the hydrogen molecule (H_2) is more stable than single atoms of hydrogen, which is why hydrogen gas is always found as molecules of two hydrogen atoms.

brass: a metal alloy principally of copper and zinc.

brazing: a form of soldering, in which brass is used as the joining metal.

brine: a solution of salt (sodium chloride) in water.

bronze: an alloy principally of copper and tin.

buffer: a chemistry term meaning a mixture of substances in solution that resists a change in the acidity or alkalinity of the solution.

capillary action: the tendency of a liquid to be sucked into small spaces, such as between objects and through narrow-pore tubes. The force to do this comes from surface tension.

catalyst: a substance that speeds up a chemical reaction but itself remains unaltered at the end of the reaction.

cathode: the positive terminal of a battery or the negative electrode of an electrolysis cell.

cathodic protection: the technique of making the object that is to be protected from corrosion into the cathode of a cell. For example, a material, such as steel, is protected by coupling it with a more reactive metal, such as magnesium. Steel forms the cathode and magnesium the anode. Zinc protects steel in the same way.

cation: a positively charged atom or group of atoms.

caustic: a substance that can cause burns if it touches the skin.

cell: a vessel containing two electrodes and an electrolyte that can act as an electrical conductor.

ceramic: a material based on clay minerals, which has been heated so that it has chemically hardened.

chalk: a pure form of calcium carbonate made of the crushed bodies of microscopic sea creatures, such as plankton and algae.

change of state: a change between one of the three states of matter, solid, liquid and gas.

chlorination: adding chlorine to a substance.

cladding: a surface sheet of material designed to protect other materials from corrosion.

clay: a microscopically small plate-like mineral that makes up the bulk of many soils. It has a sticky feel when wet.

combustion: the special case of oxidisation of a substance where a considerable amount of heat and usually light are given out. Combustion is often referred to as "burning".

compound: a chemical consisting of two or more elements chemically bonded together. Calcium atoms can combine with carbon atoms and oxygen atoms to make calcium carbonate, a compound of all three atoms.

condensation nuclei: microscopic particles of dust, salt and other materials suspended in the air, which attract water molecules.

conduction: (i) the exchange of heat (heat conduction) by contact with another object or (ii) allowing the flow of electrons (electrical conduction).

convection: the exchange of heat energy with the surroundings produced by the flow of a fluid due to being heated or cooled.

corrosion: the *slow* decay of a substance resulting from contact with gases and liquids in the environment. The term is often applied to metals. Rust is the corrosion of iron.

corrosive: a substance, either an acid or an alkali, that *rapidly* attacks a wide range of other substances.

cosmic rays: particles that fly through space and bombard all atoms on the Earth's surface. When they interact with the atmosphere they produce showers of secondary particles.

covalent bond: the most common form of strong chemical bonding, which occurs when two atoms *share* electrons.

cracking: breaking down complex molecules into simpler components. It is a term particularly used in oil refining.

crude oil: a chemical mixture of petroleum liquids. Crude oil forms the raw material for an oil refinery.

crystal: a substance that has grown freely so that it can develop external faces. Compare with crystalline, where the atoms are not free to form individual crystals and amorphous where the atoms are arranged irregularly.

crystalline: the organisation of atoms into a rigid "honeycomb-like" pattern without distinct crystal faces.

crystal systems: seven patterns or systems into which all of the world's crystals can be grouped. They are: cubic, hexagonal, rhombohedral, tetragonal, orthorhombic, monoclinic and triclinic.

cubic crystal system: groupings of crystals that look like cubes.

curie: a unit of radiation. The amount of radiation emitted by 1 g of radium each second. (The curie is equal to 37 billion becquerels.)

current: an electric current is produced by a flow of electrons through a conducting solid or ions through a conducting liquid.

decay (radioactive decay): the way that a radioactive element changes into another element because of loss of mass through radiation. For example uranium decays (changes) to lead.

decompose: to break down a substance (for example by heat or with the aid of a catalyst) into simpler components. In such a chemical reaction only one substance is involved.

dehydration: the removal of water from a substance by heating it, placing it in a dry atmosphere, or through the action of a drying agent.

density: the mass per unit volume (e.g. g/cc).

desertification: a process whereby a soil is allowed to become degraded to a state in which crops can no longer grow, i.e. desert-like. Chemical desertification is usually the result of contamination with halides because of poor irrigation practices.

detergent: a petroleum-based chemical that removes dirt.

diaphragm: a semipermeable membrane – a kind of ultra-fine mesh filter – that will allow only small ions to pass through. It is used in the electrolysis of brine.

diffusion: the slow mixing of one substance with another until the two substances are evenly mixed.

digestive tract: the system of the body that forms the pathway for food and its waste products. It begins at the mouth and includes the stomach and the intestines.

dilute acid: an acid whose concentration has been reduced by a large proportion of water.

diode: a semiconducting device that allows an electric current to flow in only one direction.

disinfectant: a chemical that kills bacteria and other microorganisms.

dissociate: to break apart. In the case of acids it means to break up forming hydrogen ions. This is an example of ionisation. Strong acids dissociate completely. Weak acids are not completely ionised and a solution of a weak acid has a relatively low concentration of hydrogen ions.

dissolve: to break down a substance in a solution without a resultant reaction.

distillation: the process of separating mixtures by condensing the vapours through cooling.

doping: adding metal atoms to a region of silicon to make it semiconducting.

dye: a coloured substance that will stick to another substance, so that both appear coloured.

electrode: a conductor that forms one terminal of a cell.

electrolysis: an electrical–chemical process that uses an electric current to cause the break up of a compound and the movement of metal ions in a solution. The process happens in many natural situations (as for example in rusting) and is also commonly used in industry for purifying (refining) metals or for plating metal objects with a fine, even metal coating.

electrolyte: a solution that conducts electricity.

electron: a tiny, negatively charged particle that is part of an atom. The flow of electrons through a solid material such as a wire produces an electric current.

electroplating: depositing a thin layer of a metal onto the surface of another substance using electrolysis.

element: a substance that cannot be decomposed into simpler substances by chemical means

emulsion: tiny droplets of one substance dispersed in another. A common oil in water emulsion is milk. The tiny droplets in an emulsion tend to come together, so another stabilising substance is often needed to wrap the particles of grease and oil in a stable coat. Soaps and detergents are such agents. Photographic film is an example of a solid emulsion.

endothermic reaction: a reaction that takes heat from the surroundings. The reaction of carbon monoxide with a metal oxide is an example.

enzyme: organic catalysts in the form of proteins in the body that speed up chemical reactions. Every living cell contains hundreds of enzymes, which ensure that the processes of life continue. Should enzymes be made inoperative, such as through mercury poisoning, then death follows.

ester: organic compounds, formed by the reaction of an alcohol with an acid, which often have a fruity taste.

evaporation: the change of state of a liquid to a gas. Evaporation happens below the boiling point and is used as a method of separating out the materials in a solution.

exothermic reaction: a reaction that gives heat to the surroundings. Many oxidation reactions, for example, give out heat.

explosive: a substance which, when a shock is applied to it, decomposes very rapidly, releasing a very large amount of heat and creating a large volume of gases as a shock wave.

extrusion: forming a shape by pushing it through a die. For example, toothpaste is extruded through the cap (die) of the toothpaste tube.

fallout: radioactive particles that reach the ground from radioactive materials in the atmosphere.

fat: semi-solid energy-rich compounds derived from plants or animals and which are made of carbon, hydrogen and oxygen. Scientists call these esters.

feldspar: a mineral consisting of sheets of aluminium silicate. This is the mineral from which the clay in soils is made.

fertile: able to provide the nutrients needed for unrestricted plant growth.

filtration: the separation of a liquid from a solid using a membrane with small holes.

fission: the breakdown of the structure of an atom, popularly called "splitting the atom" because the atom is split into approximately two other nuclei. This is different from, for example, the small change that happens when radioactivity is emitted.

fixation of nitrogen: the processes that natural organisms, such as bacteria, use to turn the nitrogen of the air into ammonium compounds.

fixing: making solid and liquid nitrogen-containing compounds from nitrogen gas. The compounds that are formed can be used as fertilisers.

fluid: able to flow; either a liquid or a gas.

fluorescent: a substance that gives out visible light when struck by invisible waves such as ultraviolet rays.

flux: a material used to make it easier for a liquid to flow. A flux dissolves metal oxides and so prevents a metal from oxidising while being heated.

foam: a substance that is sufficiently gelatinous to be able to contain bubbles of gas. The gas bulks up the substance, making it behave as though it were semi-rigid.

fossil fuels: hydrocarbon compounds that have been formed from buried plant and animal remains. High pressures and temperatures lasting over millions of years are required. The fossil fuels are coal, oil and natural gas.

fraction: a group of similar components of a mixture. In the petroleum industry the light fractions of crude oil are those with the smallest molecules, while the medium and heavy fractions have larger molecules.

free radical: a very reactive atom or group with a "spare" electron.

freezing point: the temperature at which a substance changes from a liquid to a solid. It is the same temperature as the melting point.

fuel: a concentrated form of chemical energy. The main sources of fuels (called fossil fuels because they were formed by geological processes) are coal, crude oil and natural gas. Products include methane, propane and gasoline. The fuel for stars and space vehicles is hydrogen.

fuel rods: rods of uranium or other radioactive material used as a fuel in nuclear power stations.

fuming: an unstable liquid that gives off a gas. Very concentrated acid solutions are often fuming solutions.

fungicide: any chemical that is designed to kill fungi and control the spread of fungal spores.

fusion: combining atoms to form a heavier atom.

galvanising: applying a thin zinc coating to protect another metal.

gamma rays: waves of radiation produced as the nucleus of a radioactive element rearranges itself into a tighter cluster of protons and neutrons. Gamma rays carry enough energy to damage living cells.

gangue: the unwanted material in an ore.

gas: a form of matter in which the molecules form no definite shape and are free to move about to fill any vessel they are put in.

gelatinous: a term meaning made with water. Because a gelatinous precipitate is mostly water, it is of a similar density to water and will float or lie suspended in the liquid.

gelling agent: a semi-solid jelly-like substance.

gemstone: a wide range of minerals valued by people, both as crystals (such as emerald) and as decorative stones (such as agate). There is no single chemical formula for a gemstone.

glass: a transparent silicate without any crystal growth. It has a glassy lustre and breaks with a curved fracture. Note that some minerals have all these features and are therefore natural glasses. Household glass is a synthetic silicate.

glucose: the most common of the natural sugars. It occurs as the polymer known as cellulose, the fibre in plants. Starch is also a form of glucose. The breakdown of glucose provides the energy that animals need for life.

granite: an igneous rock with a high proportion of silica (usually over 65%). It has well-developed large crystals. The largest pink, grey or white crystals are feldspar.

Greenhouse Effect: an increase of the global air temperature as a result of heat released from burning fossil fuels being absorbed by carbon dioxide in the atmosphere.

gypsum: the name for calcium sulphate. It is commonly found as Plaster of Paris and wallboards.

half-life: the time it takes for the radiation coming from a sample of a radioactive element to decrease by half.

halide: a salt of one of the halogens (fluorine, chlorine, bromine and iodine).

halite: the mineral made of sodium chloride.

halogen: one of a group of elements including chlorine, bromine, iodine and fluorine.

heat-producing: see exothermic reaction.

high explosive: a form of explosive that will only work when it receives a shock from another explosive. High explosives are much more powerful than ordinary explosives. Gunpowder is not a high explosive.

hydrate: a solid compound in crystalline form that contains molecular water. Hydrates commonly form when a solution of a soluble salt is evaporated. The water that forms part of a hydrate crystal is known as the "water of crystallization". It can usually be removed by heating, leaving an anhydrous salt.

hydration: the absorption of water by a substance. Hydrated materials are not "wet" but remain firm, apparently dry, solids. In some cases, hydration makes the substance change colour, in many other cases there is no colour change, simply a change in volume.

hydrocarbon: a compound in which only hydrogen and carbon atoms are present. Most fuels are hydrocarbons, as is the simple plastic polyethene (known as polythene).

hydrogen bond: a type of attractive force that holds one molecule to another. It is one of the weaker forms of intermolecular attractive force.

hydrothermal: a process in which hot water is involved. It is usually used in the context of rock formation because hot water and other fluids sent outwards from liquid magmas are important carriers of metals and the minerals that form gemstones.

igneous rock: a rock that has solidified from molten rock, either volcanic lava on the Earth's surface or magma deep underground. In either case the rock develops a network of interlocking crystals.

incendiary: a substance designed to cause burning.

indicator: a substance or mixture of substances that change colour with acidity or alkalinity.

inert: nonreactive.

infra-red radiation: a form of light radiation where the wavelength of the waves is slightly longer than visible light. Most heat radiation is in the infra-red band.

insoluble: a substance that will not dissolve.

ion: an atom, or group of atoms, that has gained or lost one or more electrons and so developed an electrical charge. Ions behave differently from electrically neutral atoms and molecules. They can move in an electric field, and they can also bind strongly to solvent molecules such as water. Positively charged ions are called cations; negatively charged ions are called anions. Ions carry electrical current through solutions.

ionic bond: the form of bonding that occurs between two ions when the ions have opposite charges. Sodium cations bond with chloride anions to form common salt (NaCl) when a salty solution is evaporated. Ionic bonds are strong bonds except in the presence of a solvent.

ionise: to break up neutral molecules into oppositely charged ions or to convert atoms into ions by the loss of electrons.

ionisation: a process that creates ions.

irrigation: the application of water to fields to help plants grow during times when natural rainfall is sparse.

isotope: atoms that have the same number of protons in their nucleus, but which have different masses; for example, carbon-12 and carbon-14.

latent heat: the amount of heat that is absorbed or released during the process of changing state between gas, liquid or solid. For example, heat is absorbed when a substance melts and it is released again when the substance solidifies.

latex: (the Latin word for "liquid") a suspension of small polymer particles in water. The rubber that flows from a rubber tree is a natural latex. Some synthetic polymers are made as latexes, allowing polymerisation to take place in water.

lava: the material that flows from a volcano.

limestone: a form of calcium carbonate rock that is often formed of lime mud. Most limestones are light grey and have abundant fossils.

liquid: a form of matter that has a fixed volume but no fixed shape.

lode: a deposit in which a number of veins of a metal found close together.

lustre: the shininess of a substance.

magma: the molten rock that forms a balloon-shaped chamber in the rock below a volcano. It is fed by rock moving upwards from below the crust.

marble: a form of limestone that has been "baked" while deep inside mountains. This has caused the limestone to melt and reform into small interlocking crystals, making marble harder than limestone.

mass: the amount of matter in an object. In everyday use, the word weight is often used to mean mass.

melting point: the temperature at which a substance changes state from a solid to a liquid. It is the same as freezing point.

membrane: a thin flexible sheet. A semipermeable membrane has microscopic holes of a size that will selectively allow some ions and molecules to pass through but hold others back. It thus acts as a kind of sieve.

meniscus: the curved surface of a liquid that forms when it rises in a small bore, or capillary tube. The meniscus is convex (bulges upwards) for mercury and is concave (sags downwards) for water.

metal: a substance with a lustre, the ability to conduct heat and electricity and which is not brittle.

metallic bonding: a kind of bonding in which atoms reside in a "sea" of mobile electrons. This type of bonding allows metals to be good conductors and means that they are not brittle

metamorphic rock: formed either from igneous or sedimentary rocks, by heat and or pressure. Metamorphic rocks form deep inside mountains during periods of mountain building. They result from the remelting of rocks during which process crystals are able to grow. Metamorphic rocks often show signs of banding and partial melting.

micronutrient: an element that the body requires in small amounts. Another term is trace element.

mineral: a solid substance made of just one element or chemical compound. Calcite is a mineral because it consists only of calcium carbonate, halite is a mineral because it contains only sodium chloride, quartz is a mineral because it consists of only silicon dioxide.

mineral acid: an acid that does not contain carbon and that attacks minerals. Hydrochloric, sulphuric and nitric acids are the main mineral acids.

mineral-laden: a solution close to saturation.

mixture: a material that can be separated out into two or more substances using physical means.

molecule: a group of two or more atoms held together by chemical bonds.

monoclinic system: a grouping of crystals that look like double-ended chisel blades.

monomer: a building block of a larger chain molecule ("mono" means one, "mer" means part).

mordant: any chemical that allows dyes to stick to other substances.

native metal: a pure form of a metal, not combined as a compound. Native metal is more common in poorly reactive elements than in those that are very reactive.

neutralisation: the reaction of acids and bases to produce a salt and water. The reaction causes hydrogen from the acid and hydroxide from the base to be changed to water. For

example, hydrochloric acid reacts with sodium hydroxide to form common salt and water. The term is more generally used for any reaction where the pH changes towards 7.0, which is the pH of a neutral solution.

neutron: a particle inside the nucleus of an atom that is neutral and has no charge.

noncombustible: a substance that will not burn.

noble metal: silver, gold, platinum, and mercury. These are the least reactive metals.

nuclear energy: the heat energy produced as part of the changes that take place in the core, or nucleus, of an element's atoms.

nuclear reactions: reactions that occur in the core, or nucleus of an atom.

nutrients: soluble ions that are essential to life.

octane: one of the substances contained in fuel.

ore: a rock containing enough of a useful substance to make mining it worthwhile.

organic acid: an acid containing carbon and hydrogen.

organic substance: a substance that contains carbon.

osmosis: a process where molecules of a liquid solvent move through a membrane (filter) from a region of low concentration to a region of high concentration of solute.

oxidation: a reaction in which the oxidising agent removes electrons. (Note that oxidising agents do not have to contain oxygen.)

oxide: a compound that includes oxygen and one other element.

oxidise: the process of gaining oxygen. This can be part of a controlled chemical reaction, or it can be the result of exposing a substance to the air, where oxidation (a form of corrosion) will occur slowly, perhaps over months or years.

oxidising agent: a substance that removes electrons from another substance (and therefore is itself reduced).

ozone: a form of oxygen whose molecules contain three atoms of oxygen. Ozone is regarded as a beneficial gas when high in the atmosphere because it blocks ultraviolet rays. It is a harmful gas when breathed in, so low level ozone, which is produced as part of city smog, is regarded as a form of pollution. The ozone layer is the uppermost part of the stratosphere.

pan: the name given to a shallow pond of liquid. Pans are mainly used for separating solutions by evaporation.

patina: a surface coating that develops on metals and protects them from further corrosion.

percolate: to move slowly through the pores of a rock.

period: a row in the Periodic Table.

Periodic Table: a chart organising elements by atomic number and chemical properties into groups and periods.

pesticide: any chemical that is designed to control pests (unwanted organisms) that are harmful to plants or animals.

petroleum: a natural mixture of a range of gases, liquids and solids derived from the decomposed remains of plants and animals.

pH: a measure of the hydrogen ion concentration in a liquid. Neutral is pH 7.0; numbers greater than this are alkaline, smaller numbers are acidic.

phosphor: any material that glows when energized by ultraviolet or electron beams such as in fluorescent tubes and cathode ray tubes. Phosphors, such as phosphorus, emit light after the source of excitation is cut off. This is why they glow in the dark. By contrast, fluorescors, such as fluorite, emit light only while they are being excited by ultraviolet light or an electron beam.

photon: a parcel of light energy.

photosynthesis: the process by which plants use the energy of the Sun to make the compounds they need for life. In photosynthesis, six molecules of carbon dioxide from the air combine with six molecules of water, forming one molecule of glucose (sugar) and releasing six molecules of oxygen back into the atmosphere.

pigment: any solid material used to give a liquid a colour.

placer deposit: a kind of ore body made of a sediment that contains fragments of gold ore eroded from a mother lode and transported by rivers and/ocean currents.

plastic (material): a carbon-based material consisting of long chains (polymers) of simple molecules. The word plastic is commonly restricted to synthetic polymers.

plastic (property): a material is plastic if it can be made to change shape easily. Plastic materials will remain in the new shape. (Compare with elastic, a property where a material goes back to its original shape.)

plating: adding a thin coat of one material to another to make it resistant to corrosion.

playa: a dried-up lake bed that is covered with salt deposits. From the Spanish word for beach.

poison gas: a form of gas that is used intentionally to produce widespread injury and death. (Many gases are poisonous, which is why many chemical reactions are performed in laboratory fume chambers, but they are a byproduct of a reaction and not intended to cause harm.)

polymer: a compound that is made of long chains by combining molecules (called monomers) as repeating units. ("Poly" means many, "mer" means part).

polymerisation: a chemical reaction in which large numbers of similar molecules arrange themselves into large molecules, usually long chains. This process usually happens when there is a suitable catalyst present. For example, ethene reacts to form polythene in the presence of certain catalysts.

porous: a material containing many small holes or cracks. Quite often the pores are connected, and liquids, such as water or oil, can move through them.

precious metal: silver, gold, platinum, iridium, and palladium. Each is prized for its rarity. This category is the equivalent of precious stones, or gemstones, for minerals.

precipitate: tiny solid particles formed as a result of a chemical reaction between two liquids or gases.

preservative: a substance that prevents the natural organic decay processes from occurring. Many substances can be used safely for this purpose, including sulphites and nitrogen gas.

product: a substance produced by a chemical reaction.

protein: molecules that help to build tissue and bone and therefore make new body cells. Proteins contain amino acids.

proton: a positively charged particle in the nucleus of an atom that balances out the charge of the surrounding electrons

pyrite: "mineral of fire". This name comes from the fact that pyrite (iron sulphide) will give off sparks if struck with a stone.

pyrometallurgy: refining a metal from its ore using heat. A blast furnace or smelter is the main equipment used.

radiation: the exchange of energy with the surroundings through the transmission of waves or particles of energy. Radiation is a form of energy transfer that can happen through space; no intervening medium is required (as would be the case for conduction and convection).

radioactive: a material that emits radiation or particles from the nucleus of its atoms.

radioactive decay: a change in a radioactive element due to loss of mass through radiation. For example uranium decays (changes) to lead.

radioisotope: a shortened version of the phrase radioactive isotope.

radiotracer: a radioactive isotope that is added to a stable, nonradioactive material in order to trace how it moves and its concentration.

reaction: the recombination of two substances using parts of each substance to produce new substances.

reactivity: the tendency of a substance to react with other substances. The term is most widely used in comparing the reactivity of metals. Metals are arranged in a reactivity series.

reagent: a starting material for a reaction.

recycling: the reuse of a material to save the time and energy required to extract new material from the Earth and to conserve non-renewable resources.

redox reaction: a reaction that involves reduction and oxidation.

reducing agent: a substance that gives electrons to another substance. Carbon monoxide is a reducing agent when passed over copper oxide, turning it to copper and producing carbon dioxide gas. Similarly, iron oxide is reduced to iron in a blast furnace. Sulphur dioxide is a reducing agent, used for bleaching bread.

reduction: the removal of oxygen from a substance. See also: oxidation.

refining: separating a mixture into the simpler substances of which it is made. In the case of a rock, it means the extraction of the metal that is mixed up in the rock. In the case of oil it means separating out the fractions of which it is made.

refractive index: the property of a transparent material that controls the angle at which total internal reflection will occur. The greater the refractive index, the more reflective the material will be.

resin: natural or synthetic polymers that can be moulded into solid objects or spun into thread.

rust: the corrosion of iron and steel.

saline: a solution in which most of the dissolved matter is sodium chloride (common salt).

salinisation: the concentration of salts, especially sodium chloride, in the upper layers of a soil due to poor methods of irrigation.

salts: compounds, often involving a metal, that are the reaction products of acids and bases. (Note "salt" is also the common word for sodium chloride, common salt or table salt.)

saponification: the term for a reaction between a fat and a base that produces a soap.

saturated: a state where a liquid can hold no more of a substance. If any more of the substance is added, it will not dissolve.

saturated solution: a solution that holds the maximum possible amount of dissolved material. The amount of material in solution varies with the temperature; cold solutions can hold less dissolved solid material than hot solutions. Gases are more soluble in cold liquids than hot liquids.

sediment: material that settles out at the bottom of a liquid when it is still.

semiconductor: a material of intermediate conductivity. Semiconductor devices often use silicon when they are made as part of diodes, transistors or integrated circuits.

semipermeable membrane: a thin (membrane) of material that acts as a fine sieve, allowing small molecules to pass, but holding large molecules back.

silicate: a compound containing silicon and oxygen (known as silica).

sintering: a process that happens at moderately high temperatures in some compounds. Grains begin to fuse together even through they do not melt. The most widespread example of sintering happens during the firing of clays to make ceramics.

slag: a mixture of substances that are waste products of a furnace. Most slags are composed mainly of silicates.

smelting: roasting a substance in order to extract the metal contained in it.

smog: a mixture of smoke and fog. The term is used to describe city fogs in which there is a large proportion of particulate matter (tiny pieces of carbon from exhausts) and also a high concentration of sulphur and nitrogen gases and probably ozone.

soldering: joining together two pieces of metal using solder, an alloy with a low melting point.

solid: a form of matter where a substance has a definite shape.

soluble: a substance that will readily dissolve in a solvent.

solute: the substance that dissolves in a solution (e.g. sodium chloride in salt water).

solution: a mixture of a liquid and at least one other substance (e.g. salt water). Mixtures can be separated out by physical means, for example by evaporation and cooling.

solvent: the main substance in a solution (e.g. water in salt water).

spontaneous combustion: the effect of a very reactive material beginning to oxidise very quickly and bursting into flame.

stable: able to exist without changing into another substance.

stratosphere: the part of the Earth's atmosphere that lies immediately above the region in which clouds form. It occurs between 12 and 50 km above the Earth's surface.

strong acid: an acid that has completely dissociated (ionised) in water. Mineral acids are strong acids.

sublimation: the change of a substance from solid to gas, or vica versa, without going through a liquid phase.

substance: a type of material, including mixtures.

sulphate: a compound that includes sulphur and oxygen, for example, calcium sulphate or gypsum.

sulphide: a sulphur compound that contains no oxygen.

sulphite: a sulphur compound that contains less oxygen than a sulphate.

surface tension: the force that operates on the surface of a liquid, which makes it act as though it were covered with an invisible elastic film.

suspension: tiny particles suspended in a liquid.

synthetic: does not occur naturally, but has to be manufactured.

tarnish: a coating that develops as a result of the reaction between a metal and substances in the air. The most common form of tarnishing is a very thin transparent oxide coating.

thermonuclear reactions: reactions that occur within atoms due to fusion, releasing an immensely concentrated amount of energy.

thermoplastic: a plastic that will soften, can repeatedly be moulded it into shape on heating and will set into the moulded shape as it cools.

thermoset: a plastic that will set into a moulded shape as it cools, but which cannot be made soft by reheating.

titration: a process of dripping one liquid into another in order to find out the amount needed to cause a neutral solution. An indicator is used to signal change.

toxic: poisonous enough to cause death.

translucent: almost transparent.

transmutation: the change of one element into another.

vapour: the gaseous form of a substance that is normally a liquid. For example, water vapour is the gaseous form of liquid water.

vein: a mineral deposit different from, and usually cutting across, the surrounding rocks. Most mineral and metal-bearing veins are deposits filling fractures. The veins were filled by hot, mineral-rich waters rising upwards from liquid volcanic magma. They are important sources of many metals, such as silver and gold, and also minerals such as gemstones. Veins are usually narrow, and were best suited to hand-mining. They are less exploited in the modern machine age.

viscous: slow moving, syrupy. A liquid that has a low viscosity is said to be mobile.

vitreous: glass-like.

volatile: readily forms a gas.

vulcanisation: forming cross-links between polymer chains to increase the strength of the whole polymer. Rubbers are vulcanised using sulphur when making tyres and other strong materials.

weak acid: an acid that has only partly dissociated (ionised) in water. Most organic acids are weak acids.

weather: a term used by Earth scientists and derived from "weathering", meaning to react with water and gases of the environment.

weathering: the slow natural processes that break down rocks and reduce them to small fragments either by mechanical or chemical means.

welding: fusing two pieces of metal together using heat.

X-rays: a form of very short wave radiation.

Index